NEXTGEN LEARNERS PRESENTS

Think Like a Brain!™ Collection

Tammy Temporal & The Soundy Sound Storm

Understanding Sounds, Words, and What You Hear

BOOK 5

Think Like a Brain!™
Collection Overview

Think Like a Brain!™ is a wildly engaging, silly-smart children's book collection that invites kids to explore how their brains work from the inside out. Through playful characters, imaginative settings, and rhythm-rich storytelling, this collection teaches neuroscience, social-emotional learning (SEL), and self-understanding in ways children absorb without even realizing they are learning.

Instead of telling children how to behave or what to feel, Think Like a Brain!™ shows them why their brains respond the way they do. Each story transforms complex brain science into friendly characters and relatable moments, helping kids feel capable, curious, and empowered rather than confused or ashamed.

Designed for children first and embraced by parents and educators as preventative education, this collection supports:

- Emotional regulation and self-awareness
- Executive functioning skills
- Neurodivergent understanding and inclusion
- Brain-body connection and growth
- Compassion for self and others

With humor, heart, and factual accuracy at its core, Think Like a Brain!™ helps children build the skills many adults wish they had learned sooner, one playful page at a time.

One of the collection's most powerful features is how the characters become real-life reminders for children long after the story ends. Once kids learn who the characters are and what jobs they do, they can begin to recognize those same processes in themselves. A big feeling might become, "It looks like Amy is trying to tell you something," or a moment of pause might sound like, "What does your Fred think?" By naming experiences through familiar characters, children learn to notice what is happening inside their brains, reflect on it, and move forward more quickly with understanding rather than frustration.

Series 1:

The Brainy Bunch (Brain Anatomy)

Series Overview: Meet the brain's core parts as a friendly team. Kids learn what each part does and how they work together to help thinking, feeling, moving, and learning.

1. **The Brainy Bunch**
   Focus: Brain overview and teamwork
2. **Brainstem Benny & The Automatic Adventure**
   Focus: Breathing, heartbeat, survival basics
3. **Bella Cerebellum's Balance Ballet**
   Focus: Balance, coordination, practice
4. **Olly Occipital & The Lookout Tower**
   Focus: Vision, visual processing
5. **Tammy Temporal & The Soundy Sound Storm**
   Focus: Hearing, language, sound processing
6. **Perry the Parietal & The Body Map Bash**
   Focus: Touch, space, body awareness
7. **Hippocampus Hal & The Memory Maze**
   Focus: Memory, learning, recall
8. **Amy the Amygdala & The Overactive Alarm**
   Focus: Emotional alarm system
9. **Luna Lefty & Roger Righty's Mixed Up Masterpiece**
   Focus: Left/right hemispheres, teamwork
10. **Callie the Connector & The Brain Bridge**
    Focus: Corpus callosum, brain-to-brain communication

Series 2:
Cortex Castle (Frontal Lobe Executive Functions)

Series Overview: A lively castle of executive-function helpers. Kids practice planning, starting, stopping, focusing, flexible thinking, time skills, and self-talk, with Frontal Fred guiding thoughtful choices.

1. **Cortex Castle**
   <u>Focus</u>: Executive function overview
2. **Frontal Fred & The Decision Dilemma**
   <u>Focus</u>: Decision-making, prioritizing
3. **Impulsive Ian & The Button Box**
   <u>Focus</u>: Impulse control, pause
4. **Memory Molly & The Wandering Thoughts**
   <u>Focus</u>: Working memory, holding steps
5. **Focus Frannie & The Great Distraction Chase**
   <u>Focus</u>: Attention, refocusing
6. **Planella & The Checklist Parade**
   <u>Focus</u>: Planning, sequencing
7. **Flexi & The Changey-Changeroo Day**
   <u>Focus</u>: Cognitive flexibility
8. **Self-Talk Sally & The Thoughtful Echoes**
   <u>Focus</u>: Self-talk, emotional regulation
9. **Starter Stacy & Finisher Finn**
   <u>Focus</u>: Task initiation, task completion
10. M**ilo the Minute Keeper & Mae the Moment Seeker**
    <u>Focus</u>: Time awareness, pacing
11. **The Power of Pausing with Penny**
    <u>Focus</u>: Pause skills, regulation

Series 3:
Busy Brain Café (Neurodivergent Affirmations)

Series Overview: A welcoming café where different brain styles are celebrated. These stories affirm neurodivergent experiences while teaching self-understanding, communication, and compassion for others.

1. **Nathan Neuron & The Bouncy Brain**
   Focus: ADHD metaphor, movement, attention
2. **Sara Sprinkle the Sensa & The Super Sensitive Day**
   Focus: Sensory sensitivity, overload tools
3. **Overwhelmed Owen & The Senses Overload**
   Focus: Sensory overload, coping strategies
4. **Interoceptive Ivy & The Inside Signals**
   Focus: Interoception, body signals
5. **Zane's Zigzag Thoughts & The Path to Peace**
   Focus: Anxiety, grounding
6. **The Busy Brain Café**
   Focus: Neurodivergent inclusion, different processing styles
7. **The Listening Light**
   Focus: Nonverbal communication, understanding
8. **The Language of Love**
   Focus: Connection, communication, compassion
9. **Trying Theo & The How-To Hiccup**
   Focus: Persistence, frustration tolerance
10. **Drowsy Drew & The Nighttime Repair**
    Focus: Sleep, brain repair
11. **Burned-Out Bonnie & The Rebuild**
    Focus: Burnout, rest, recovery

Series 4:
Steady Systems Squad (Regulation & Growth)

Series Overview: A squad that explains how brains regulate, reset, rest, and grow stronger over time, including stress waves, sleep crews, sensory filtering, and rebuilding after hard days.

1. **The Steady Systems Squad**
   Focus: Regulation systems overview
2. **Timothy Thalamus & The Traffic Tower**
   Focus: Sensory filtering, attention routing
3. **Danny Dimmer & Wake-Up Wonda**
   Focus: Arousal and alertness
4. **The Start, Stop, and Switch Station**
   Focus: Initiation, inhibition, shifting
5. **The Night Shift & The Sleepy Crew**
   Focus: Sleep systems, restoration
6. **The Stress Wave & The Return to Calm**
   Focus: Stress response, recovery
7. **The Brain Growth Gathering**
   Focus: Brain growth and development
8. **Paul's Practice Path**
   Focus: Neuroplasticity
9. **Daisy's Daydream Network**
   Focus: Daydream network, imagination
10. **Remy's Rebuild Crew**
    Focus: Recovery and healing

NEXTGEN LEARNERS PRESENTS

Think Like a Brain!™ Collection

Tammy Temporal & The Soundy Sound Storm

Understanding Sounds, Words, and What You Hear

BOOK 5

WRITTEN &
ILLUSTRATED BY:

MAISEL MCLAULA

NEXTGEN LEARNERS PRESS™
CINCINNATI, OHIO

This book is published by NextGen Learners Press, an imprint of NextGen Learners LLC.

Maisel McLaula is a pen name used by the author of this work.

First Edition: 2026
Printed in the United States of America
ISBN: 978-1-972860-04-5

This book is intended for educational and informational purposes only. It is not intended to replace professional medical, psychological, or therapeutic advice, diagnosis, or treatment.
Always seek the advice of a qualified professional with any questions regarding a medical or mental health condition.

Artwork and visual elements in this book were created using a combination of original artwork, with enhanced original direction, digital illustration tools, and AI-assisted creative technologies.

A Note for My Readers:

Hi friend,
Your brain is incredible. It thinks, feels, remembers, solves, dances, creates, and so much more!
This series is here to help you understand how your brain works, and to cheer you on as you grow your own thinking superpowers.

Every part of your brain has a job.
Sometimes they get mixed up. That's okay!

When you learn what each part does, you can work with your brain instead of feeling like it's working against you.
So read, giggle, learn, and maybe even draw your own Brainy Bunch when you're done.

I'm cheering you on with every rhyme.

-Maisel McLaula

THINK LIKE A BRAIN!™

BRAIN ANATOMY COLLECTION

EXPLORE THE CASTLE INSIDE YOUR HEAD! MEET EACH ROOM AND ITS BRILLIANT RESIDENT, FROM FRONTAL FRED TO BENNY THE BRAINSTEM. WHIMSY, WONDER, AND NEURONS AWAIT!

For Keely,
with patience and love, may you always remember there is space to pause, breathe, and feel safe as you listen.

And...

To the beautifully wired, the curiously wired, and the wonderfully neurodivergent minds, may you find the volume that feels right for you.

You are exactly what a great brain looks like.

-Maisel McLaula

In the Echo Hallway where noises bounced around,
Lived Tammy Temporal, ready for sound.
She heard every whisper, each murmur and squeak,
A crackle, a cackle and crows that could speak!

She listened to music, to raindrops and the tick-tock of time,
She kept perfect tempo and marched in rhythm and rhyme.

She translated voices and picked up each tone,
Even sighs that you made when you thought you were alone.

But one thunderous day, when the hall got too loud,
With clatter and music and a chattery crowd,

The sounds swirled around her like wind in a jar,
And clouds of fury were thick from afar.

"Too many noises!" she gasped with a frown,
As bells and buzzers came raining down.

"Was that someone talking? Or a cat saying moo?
My brain's all tangled, I don't know what to do!"

Fred tiptoed in with some earmuffs and books,
He brought gentle noises and comforting looks.

“We need a calm corner, a break from the sound,
Let’s help you feel safe while you sort what you’ve found.”

They built a Sound Space with cushions and zones,
Soft humming pillows and musical tones.

They practiced with rhythms, with beats slow and clear,
So Tammy could focus and feel what she'd hear.

She labeled the noises: a door creak, a drop,
A bird's gentle tweet, a whistle for stop.

She mapped every pitch, every volume and beat,
And danced through the echoes with feather-light feet.

With each sound she named, the storm grew more still,
She found she had space to both listen and feel.

Her brain could now sort what it used to all blend,
And her tangled-up sounds made a song in the end.

Now Tammy still hears all the highs and the lows,
But she filters and chooses where her attention goes.

She uses her tools when the noise gets too wide,
And she always keeps quiet time tucked by her side.

She helps all her friends when the sound feels too strong,
By playing soft music or changing the song.

She knows that for some, loud is just too much,
And a whisper or hum can be just the right touch.

Tammy now teaches that hearing is more,
Than just catching noise when you walk through the door.

It’s noticing the tone, understanding the beat,
And choosing the rhythm that feels kind and sweet.

So if sound feels like thunder or clatter or boom,
And you feel like your ears take up too much room,

Try Tammy's soft space, or tap out a song,
You can listen with power and still get along.

Tammy's Sound Tools

- Make a quiet space for your ears to rest
- Use rhythm and movement to sort what is best
- Try humming or tapping to filter the noise
- Label each sound with intention and poise

What Is the Temporal Lobe?

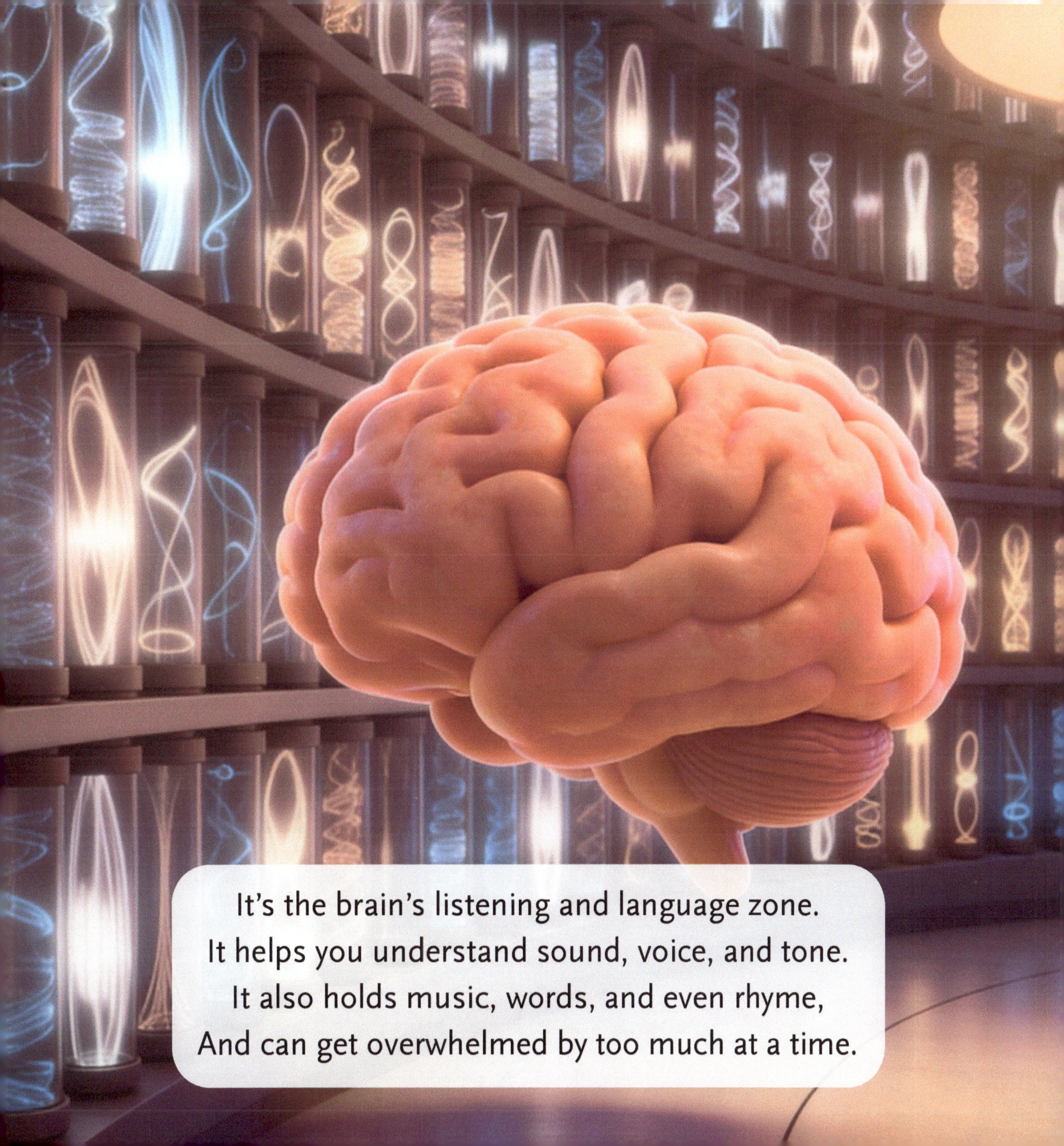

It's the brain's listening and language zone.
It helps you understand sound, voice, and tone.
It also holds music, words, and even rhyme,
And can get overwhelmed by too much at a time.

Try It Yourself!

- Create a Sound Space where listening's allowed
- Then sort what you hear: quiet, medium, or loud
- Clap to a beat or hum like a bee
- And see what your brain hears more easily

Congratulations!

You've heard with Tammy and tamed the sound storm,
You've learned how to filter and soften the swarm.
With rhythm, with rhyme, with space when you need,
You can listen with care and let your thoughts lead.

Thank You to My Readers

Thank you for reading this book.
Whether you're a parent, a teacher,
a therapist, or a brilliant young
brain explorer,
your interest in understanding
how the mind works
means the world.

Sharing these books
means spreading compassion and
curiosity.

You are the reason
these stories come alive.
Thank you for believing in the
power of playful learning.

Stay curious. Stay kind.
Stay brainy!

Tammy Temporal & The Soundy Sound Storm

Teacher & Parent Guide

💡 What This Story Teaches:

- Brain part: Temporal Lobe (sound, language, rhythm, meaning)

- Core idea: Hearing is more than ears. The brain must organize sound.

- Skills supported: auditory processing, language understanding, rhythm, listening stamina

- Big message: Too much noise can overwhelm the brain. Calm sound helps thinking return.

Tammy teaches children that listening is brain work, and that quiet, rhythm, and pauses help sounds make sense.

🧠 Key Concepts for Grown-Ups

Tammy represents the temporal lobes, located on the sides of the brain. These regions help process sound, spoken language, music, tone of voice, and rhythm. They also support memory for words and meaning.

When auditory input becomes overwhelming, children may show:

- Covering ears
- Difficulty following directions
- Meltdowns in noisy environments
- Trouble understanding speech in groups
- Fatigue after busy days
- Sensitivity to sudden sounds

Important foundations:

- Auditory processing requires sorting, not just hearing.
- Background noise competes with learning.
- Rhythm helps organize sound.
- Quiet moments restore listening capacity.

This story helps children understand that:

- Feeling overwhelmed by noise is not misbehavior.
- The brain needs breaks from sound.
- Calm listening can be practiced.

Discussion Starters

- What sounds made Tammy feel overwhelmed?
- What helped Tammy calm the sound storm?
- Which sounds feel nice to you?
- Which sounds feel too big?
- How can we help our ears rest?

Encourage children to name sounds without judgment.

👪 Try This Together:

1. Sound Sorting

Listen to nearby noises and sort them into quiet, medium, or loud.

2. Rhythm Reset

Clap or tap a slow steady beat together for 30-60 seconds.

3. Humming Calm

Hum softly or sing one gentle note to help regulate breathing and sound.

4. Quiet Corners

Create a small listening break area with pillows or soft textures.

5. Voice Levels

Practice whisper voice, talking voice, and outdoor voice.

Bonus Challenge

Build Tammy's Sound Space

Invite children to draw or build a cozy listening space.

Ask:

- What helps Tammy hear clearly?
- What sounds belong in the calm zone?
- What tools help when noise feels too big?

Optional Labels:

- Quiet Button
- Rhythm Pad
- Humming Spot
- Soft Sound Station

Playful Extension:

Sound Detective

Take turns making gentle sounds while others guess what made the noise.

This builds focused listening without overwhelm.

🛠️ Tammy's Sound Toolbox:

Use these tools when children feel auditory overload:

- Reduce background noise when possible
- Use visual cues alongside verbal directions
- Speak slowly and clearly
- Provide headphones or ear defenders
- Add rhythmic movement
- Schedule quiet recovery time
- Offer humming or soft music

Helpful phrases:

- "Let's give your ears a break."
- "Your temporal lobe is sorting sounds."
- "We can slow the noise together."

Avoid giving important instructions in noisy spaces.

❤️ Final Note for Grown-Ups

Tammy reminds us that listening is a full-body experience.

When sound feels overwhelming, children are not refusing to cooperate. Their brains are asking for quieter input.

With rhythm, pauses, and gentle sound, children learn that they can calm the storm and hear clearly again.

Listening grows best in safety and softness.

Resources & GROWTH

Resources:

(Temporal lobe + hearing, sound processing, language)

- National Institute on Deafness and Other Communication Disorders NIDCD. How Do We Hear? (auditory pathway to the brain).
- Cleveland Clinic. Hearing: how the auditory process works (includes brainstem → temporal lobe processing).
- National Institute of Neurological Disorders and Stroke (NINDS). Brain Basics: Know Your Brain / Brain Basics resources.
- Cleveland Clinic. Brain: parts and function (overview pages).
- CASEL. CASEL's SEL Framework (self-awareness, self-management, social awareness, relationship skills, responsible decision-making).
- CDC. Child Development: Positive Parenting Tips and developmental guidance by age.
- MedlinePlus (U.S. National Library of Medicine)
    - Search: auditory processing
- Encyclopedia Britannica
    - Look up: temporal lobe

Growth:

(If you or someone you know wants to build sound tolerance or listening skills)

- American Speech-Language-Hearing Association (ASHA)
    - Look for: auditory processing
- Child Mind Institute
    - Search: sound sensitivity

Acknowledgments:

To my students: past, present, and future:
You taught me more than any textbook ever could. By watching how you explored, questioned, struggled, and discovered, I learned how minds truly grow. Every "aha" moment in these pages began with you.

To the parents, teachers, and guides walking beside children every day:
Thank you for listening, for advocating when things feel complicated, and for believing that understanding a child's inner world truly matters. Your patience and care shape futures in ways you may never fully see.

To my husband, my best friend, and my constant partner in learning:
Thank you for walking beside me, believing in this work, and holding space for creativity to exist. Sharing curiosity, growth, and creation with you has been one of the greatest honors of my life.

To the great thinkers who shaped how I see the world: including Dr. Seuss, Albert Einstein, Leonardo da Vinci, Stephen Hawking, Ludwig van Beethoven, just to name a few, and my English teacher, Mrs. Godfrey:
Your curiosity, courage, and imagination reminded me that learning is limitless, creativity is powerful, and questions matter. Mrs. Godfrey, your encouragement, belief in me, and love of words sparked my writing journey. If not for you, I may never have picked up a pen.

And to my publisher, NextGen Learners™:
It was by fate that we met, and by inspiration that we now move forward. Together, we build something meaningful. Together, we leave this legacy.

Lastly, to my own brain:
Complex, resilient, ever-adapting, and endlessly learning. Thank you for carrying stories, holding wonder, and reminding me that growth is always possible.

Maisel McLaula is the author of the Think Like a Brain!™ children's book collection, a story-first series that blends neuroscience, imagination, and emotional safety for young readers.

Drawing from decades of experience supporting children and families, her work centers on a simple but powerful belief: children deserve to understand and accept who they are, including how their brains work and why they make the choices they do.

Through playful characters and gentle metaphors, Maisel's stories help children discover the "why" behind their thoughts, feelings, and actions, turning confusion into clarity and self-judgment into self-compassion. By understanding their inner systems, children gain the power to respond with kindness toward themselves and to recognize those same patterns in others.

Maisel believes that when children feel safe inside their own minds, empathy grows naturally. The Think Like a Brain!™ collection offers a foundation for confidence, understanding, and care that grows with the child.

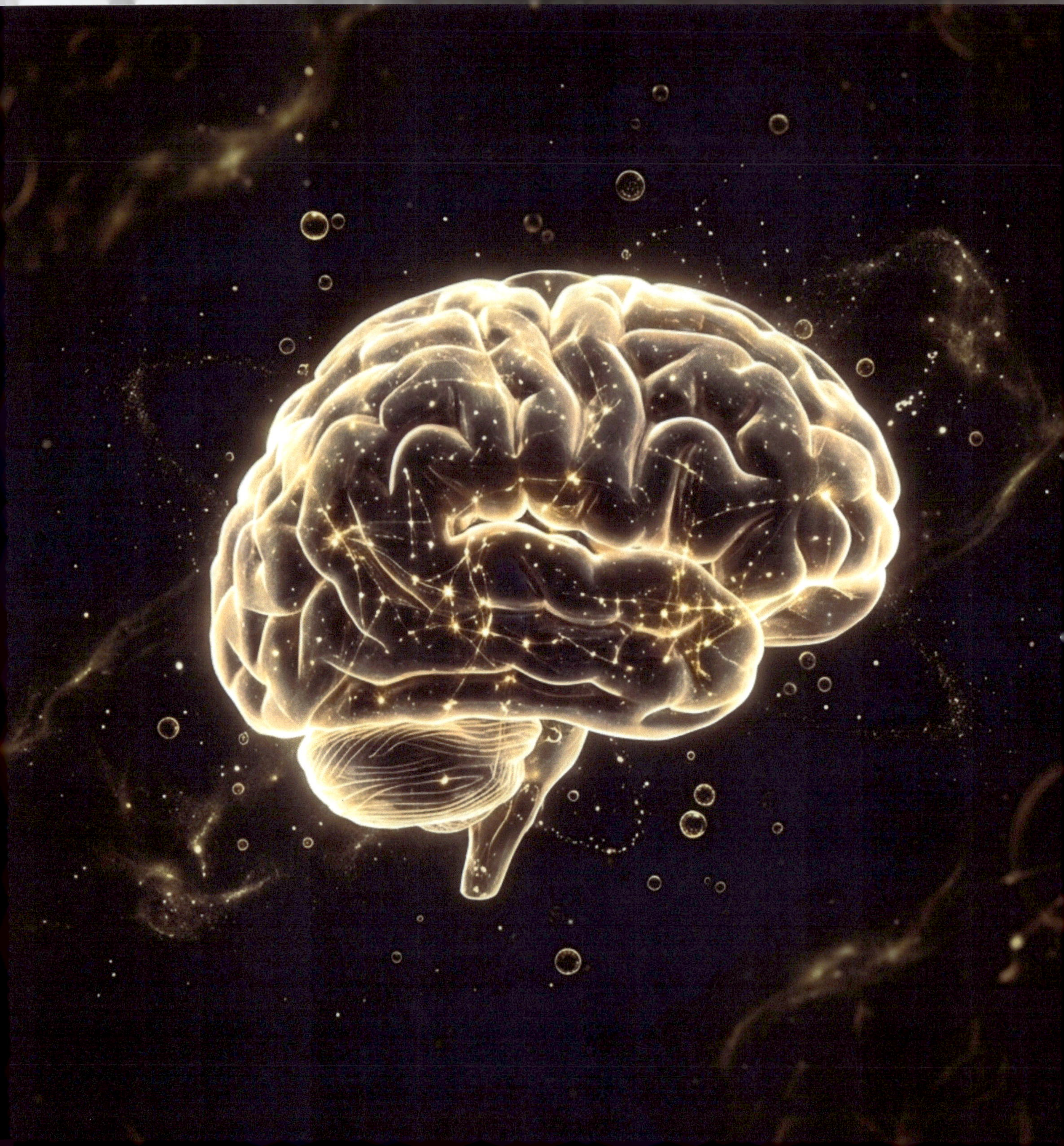

ABOUT THE PUBLISHER

NextGen Learners LLC is an educational publishing company dedicated to helping children understand how they think, feel, learn, and grow.

Through imaginative storytelling, neuroscience-informed concepts, and real-world application, NextGen Learners™ creates books and resources that support emotional development, critical thinking, self-awareness, and confidence in young minds.

Our work is designed not only for children, but for the parents, educators, caregivers, and professionals who walk beside them. Each book serves as a bridge, helping adults better understand the child in front of them, while helping children better understand themselves.

By creating shared language and meaningful connection points, our stories foster deeper relationships, reduce misunderstanding, and encourage acceptance, both within oneself and toward others.

We believe every child's mind is uniquely wired, and that understanding those differences is the foundation for compassion, growth, and lifelong learning.
With multiple collections designed to support cognitive, emotional, and developmental growth, NextGen Learners provides tools that children can carry with them, and tools that adults can use to guide them with clarity, patience, and purpose.

NG
NEXTGEN
LEARNERS
PRESS

MORE TO EXPLORE

The Brainy Bunch is just the beginning.
More stories, tools, and brain-friendly resources are on the way! Designed to help children, parents, and educators better understand how the brain works in everyday life.

FREE PRINTABLES

We've created printable pages to go along with this book, including:

-Reading Guides
-Brain helper visuals
-Character Cards

These are available for free as part of our growing resource library.

Our full website experience is built to include:

-Downloadable activity packs
-Parent & teacher guides
-Expanded learning tools
-Additional collections and topics

STAY CONNECTED

We love hearing how children connect with these stories!

If there is a topic, feeling, or challenge you would like to see explored in future books, we would love to hear from you!

Contact us: Inspire@NextGenLearners.com
NextGenLearners.com

Your ideas help shape future stories and support the growth of this collection.

This collection grows with you. New resources and stories are added regularly.

Maisel McLaula loves hearing from the tiny humans who read her books.

Did a story make you giggle, spark a big idea, or inspire you to draw your very own character? She would be delighted to see it! Send your letters and artwork to:

NextGen Learners
Attn: Maisel McLaula
8666 Beechmont Ave #1063
Cincinnati, OH 45255

Every letter is read with a smile, and your artwork just might make her whole day.

OPEN BRAIN

www.ingramcontent.com/pod-product-compliance
Lightning Source LLC
Chambersburg PA
CBHW041322210926
51847CB00002B/28